GET INFORMED—STAY INFORMED
PLASTICS DEPENDENCY

Natalie Hyde

CRABTREE
PUBLISHING COMPANY
WWW.CRABTREEBOOKS.COM

Author: Natalie Hyde
Series research and development:
  Reagan Miller
Editor-in-chief: Lionel Bender
Editors: Simon Adams, Ellen Rodger
Proofreaders: Laura Booth,
  Crystal Sikkens
Project coordinator: Melissa Boyce
Design and photo research:
  Ben White
Production: Kim Richardson
**Production coordinator and
  Prepress technician:** Margaret Amy Salter
Print coordinator: Katherine Berti
Consultant: Emily Drew,
  Public Librarian, B.F.A., M.S.-LIS

Produced for Crabtree
Publishing Company by
Bender Richardson White

Photographs and reproductions: A Plastic Ocean Movie, 19 (A Plastic Ocean Foundation (www.aplasticocean.foundation) rightful owner of award winning documentary A Plastic Ocean); Alamy: 12–13 (Granger Historical Picture Archive), 38–39 (Xinhua); Getty Images: 10 (Frederick M. Brown/Stringer), 18 (SOPA Images), 30–31 (Romeo Gacad), 32 (Smith Collection/Gado), 33 (Anadolu Agency), 34–35 (Tang Chhin Sothy), 41 (Andrew Lichtenstein); MacRebur: 21 (Clay10 Creative for MacRebur); Science Photo Library: 17 (Steve Gschmeissner), 37 (Patrice Latron/Eurelios/Look at Sciences); Shutterstock: 1 (Larina Marina), 4–5 (vovidzha), 6–7 (Radu Bercan), 7 (Flamingo Images), 8–9 (dimitris_k), 10–11 (Rawpixel.com), 14–15 (ARENA Creative), 15 (bondvit), 16–17 (Rich Carey), 22–23 (Alfa Photostudio), 24 (Robert Gregory Griffeth), 24–25 (Pix One), 26–27 (Kev Gregory), 26–27 (Maxim Blinkov), 28–29 (Gigira), 29 (RecycleMan), 35 (Albert Pego), 36 (dcurzon), 40–41 (Casimiro PT), 40–41 (aSuruwataRi), 42–43 (Monkey Business Images), 43 (Igisheva Maria); Icons & heading band: shutterstock.com

Diagrams: Stefan Chabluk, using the following as sources of data: p. 7 Geyer et al. (2017), Science Advances mag 3(7). p. 14 Geyer et al. (2017), Science Advances mag 3(7). p. 17 Statista.com/The Wall Street Journal. p. 20 ASTM International/Roland Geyer, Univ. of California. p. 24 Statista.com/Eurostat. p.26 PlasticsEurope/Population Reference Bureau/Nat. Geo. magazine. p. 29 Deutsche Welle/Eurostat. p.32 National Oceanic and Atmospheric Administration, U.S.A.

Library and Archives Canada Cataloguing in Publication

Title: Plastics dependency / Natalie Hyde.
Names: Hyde, Natalie, 1963- author.
Series: Get informed--stay informed.
Description: Series statement: Get informed, stay informed |
  Includes bibliographical references and index.
Identifiers: Canadiana (print) 20190236884 |
  Canadiana (ebook) 20190236892 |
  ISBN 9780778772743 (hardcover) |
  ISBN 9780778772798 (softcover) |
  ISBN 9781427124678 (HTML)
Subjects: LCSH: Plastics—Environmental aspects—
  Juvenile literature. | LCSH: Plastic scrap—Environmental
  aspects—Juvenile literature. | LCSH: Plastics industry and trade—
  Social aspects—Juvenile literature.
Classification: LCC TP1125 .H93 2020 | DDC j668.4—dc23

Library of Congress Cataloging-in-Publication Data

CIP available at the Library of Congress

LCCN: 2019053902

Crabtree Publishing Company
www.crabtreebooks.com   1-800-387-7650

Printed in the U.S.A./032020/CG20200127

Published in Canada
Crabtree Publishing
616 Welland Ave.
St. Catharines, ON
L2M 5V6

Published in the United States
Crabtree Publishing
PMB 59051
350 Fifth Avenue, 59th Floor
New York, NY 10118

Published in the United Kingdom
Crabtree Publishing
Maritime House
Basin Road North, Hove
BN41 1WR

Published in Australia
Crabtree Publishing
Unit 3 – 5 Currumbin
Court
Capalaba
QLD 4157

CONTENTS

1 THE NEED TO KNOW

In the United States alone, people throw away enough plastic water bottles each day to completely circle Earth. Around the world, we use about 160,000 plastic bags each second! Scientists predict that by 2050 there may be more plastic in the oceans than fish.

> *Plastics are ... pretty well everywhere on Earth, from mountain tops to the deep ocean floor.*
>
> Jan Zalasiewicz, British geologist, 2016

▶ Seventy-three percent of beach litter around the world is plastic, including bottles, bottle caps, and food wrappers.

QUESTIONS TO ASK

Within this book are three types of boxes with questions to help your critical thinking about our use of plastics. The icons will help you identify them.

THE CENTRAL ISSUES
Learning about the main points of information.

WHAT'S AT STAKE?
Helping you determine how the issue will affect you.

ASK YOUR OWN QUESTIONS
Prompts to address gaps in your understanding.

GOING NOWHERE

The problem is, plastics do not break down like natural materials such as wood, leather, and cotton. Once created, plastic remains as it is made for many thousands of years. **Recycling** used plastic allows us to reuse it as new products, but not every type of plastic is easily recycled. Even if a recycling process exists, sometimes it is too expensive to use or there is no market for the recycled plastic product. This means more plastic is being sent to **landfills**. Once there, it will sit in the ground for hundreds or even thousands of years. As Earth's population grows, and we use and throw away ever more plastic, it is piling up in our landfills, clogging our oceans, and finding its way into our **food chain** and our bodies.

A PART TO PLAY

We all have a part to play in the problem of plastic **pollution**. Understanding how plastics are made, what happens to plastic items when we are done using them, and the dangers plastic poses to us and our **environment** is vital to our health and the health of our planet. Because plastic products are **convenient** and cheap to make and use, our **dependence** on plastics has led to worldwide problems. A solution is not easy or straightforward. With good and complete information, we can make good decisions about how to fight for change. That might mean holding companies and governments responsible. It might also mean a change in our own habits of buying and using plastics in our lives.

Plastics are found everywhere in our lives. They are in items for cooking, for shopping, in our schools, in our clothes, in our electronics, and even in our bathrooms. Getting rid of plastics completely would affect every part of our lives.

Things that used to be made of wood, metal, or plant **fibers** are now often made of plastic. Part of the reason plastic items are so popular is because plastic is waterproof, strong, **durable**, and cheap to buy. Plastic is lightweight, which means shipping plastic items from where they are manufactured overseas is cheaper than shipping something made of heavier metal or wood.

THE PROBLEMS AND THE DEBATE

The qualities that make plastic desirable are also the ones that pose risks. Plastics do not break down quickly. Their cheap price means more industries want to use plastics to reduce costs, and we want to pay less when buying products. Tiny particles of plastic, called **microplastics**, are showing up everywhere on the planet, including in our drinking water and the food we eat. The microplastics then end up in our own bodies, with consequences that are yet to be identified.

The plastics dependency **debate** centers on who is responsible for the risks and damage from plastics to our health and the health of our world; what we should do about the amount of plastic showing up in every corner of our planet; and who should work to fix the problem. The companies producing plastics are facing very little pressure to cut back their production. To solve the problem, we first need to understand it and get a balanced view by seeing all sides of the issue. With good and correct information, we can create our own personal strategy to stay informed.

ASK YOUR OWN QUESTIONS

To determine if a source is credible, consider:

- Does the creator have solid credentials and expertise in the topic?
- Does the headline match the story?
- Is the publisher known to be reliable?
- What **source materials** did the creator use?
- Is the source you're studying up to date?
- Is the source meant to be a joke or **clickbait**?

▶ Each one of us uses an average of 11 bottles of shower gel and 10 bottles of shampoo every year. Almost all these plastic containers are easily recycled.

MAJOR USERS OF PLASTICS, BY ECONOMIC SECTOR
Based on metric tons of primary plastic produced globally in 2015

Electrical/Electronic **4.5%**

1% Industrial Machinery

Transportation

Homeware, Schools, Offices

6.5%

10%

Packaging

37%

11%

Textiles

14%

16%

Toys, Sports, Leisure

Building and Construction

▲ The Internet can be a valuable source of information providing you are careful to check the facts you read.

It is sometimes **daunting** to begin to learn about a new topic. There can be an overwhelming amount of information that is spread out in many places. There might also be a lot of different opinions from people you trust. How do you sort it all out? Where do you start?

▶ Our electronics contain about 17 percent plastic. The type of plastic used is easy to recycle and reuse.

8

KEY PLAYERS

The **U.S.** and **Canadian plastic industry associations** are organizations that want to be a voice for plastics sustainability in North America. They are dedicated to the growth of the plastics industry. Reports and articles focus on a positive image for the industry. Their websites describe programs to clean up plastics in our oceans, new ways of recycling, and new plastics technologies.

TAKING THE FIRST STEP

The first step to getting an insight into our plastics dependency is to get familiar with the key players, vocabulary, and background to the plastics issue. Key players will be petroleum producers, industrialists, recyclers, environmentalists, and us, the **consumers**. Key vocabulary will include the many different chemicals used to make plastics, the various recycling terms, and new technology. A timeline showing how and when plastics were developed can give you an overview of why we are so dependent on plastics today.

SEEING BIAS

As you start to gather information, you will notice that some reports and articles have a slant toward a certain set of beliefs. Some may indicate that plastics are a necessary, important, and useful material that benefits our lives. Others may show that our world and our health is being damaged by tiny plastic particles and chemicals **leaching** out of plastics. This slant or focus is called **bias**, and it is our personal opinion in favor of or against something or someone. Everyone has their own bias. Bias is not necessarily bad, but it is important to recognize it when you are researching a topic. Scientific and government reports strive to be more objective and eliminate any bias.

> *Plastic waste does not only pollute our oceans. Burning plastic in incinerators turns one form of pollution into another, whether it be air emissions, **toxic** ash, or wastewater.*

Doun Moon, Research Associate, GAIA (Global Alliance for Incinerator Alternatives), 2019

▲ British wildlife broadcaster Sir David Attenborough has described plastic pollution as an "unfolding catastrophe." His *Blue Planet II* TV series focuses on the issue surrounding plastic waste.

▶ Libraries not only have a lot of information in print books and magazines, but they also have Internet access available for people who do not have easy access at home.

KEY INFORMATION

Primary sources are the original creators or owners of information, for example a report on the amount of microplastics in the ocean. **Secondary sources** are reports, analyses, and **interpretations** of the primary sources, for example, a magazine article using the microplastics report to estimate how long ocean clean-up would take. **Tertiary sources** are summaries or databases of primary and secondary information. They include Wikipedia articles or entries in encyclopedias.

Information on a topic is called source material. It can be written documents, images, or audio files. It can be created by manufacturers, scientists, businesses, health professionals, and consumers. Source material is fairly easy to find, but can be spread out in many different places. Gathering, sorting, and reading it is part of getting informed.

WHERE TO LOOK

Source materials about plastics can be found in libraries, **archives**, and on the Internet. Any industry or organization has to keep records. Petroleum producers, plastic manufacturers, landfill management, and recyclers all create records of their programs and profits. These reports and documents are often made public.

Governments collect **statistics** and data in order to enforce laws and regulations about plastics. Newspapers and magazines, such as *The New York Times* and the *Globe and Mail*, publish articles and stories about plastics and how they are **impacting** our environment. News outlets, either on TV or online, report on any new happenings. Scientific journals on how plastics are made or how and when they break down are in libraries or online. Social media is a good place to find mentions of articles on discoveries and to gather a variety of viewpoints on a topic such as plastics.

When you are getting informed, ask yourself what might **motivate** a person or group to have you believe one thing over another. It is also important to keep the **context** of the source material in mind. Context refers to the time, setting, or situation information is created in. An article written when plastics were first developed might have a more positive slant than an article written today as landfills are filling up with waste.

Plastics are made of a group of materials called **polymers**. Polymer means "of many parts." Polymers can be found in nature such as the fiber in plants called **cellulose**. Polymers can also be created in a laboratory using either natural substances or **synthetic** ones, like long chains of carbon atoms found in petroleum and fossil fuels.

▶ Before plastic was widely available, stores packaged goods in paper bags, burlap sacks, and cardboard boxes. These materials all decompose easily in landfills.

TIMELINE

1869 First commercial plastic made to replace ivory used for billiard balls
1889 Eastman Kodak invents **celluloid** film
1907 Bakelite is invented
1931 **Vinyl** records introduced by RCA
1938 **Nylon** first used as bristles in toothbrushes
1950 Polyester is created
1965 Kevlar developed and used in bulletproof vests
1977 First plastic bags offered at stores in the U.S.
1988 First polymer banknote issued in Australia
2002 Ireland becomes the first country to charge for plastic bags
2016 California becomes the first U.S. state to ban single-use plastic bags
2018 Prince Edward Island becomes the first Canadian province to ban plastic bags

> We are told that 16 percent of all the animals that are found dead on the coast are dead as a result of their interaction with plastic bags. The plastic bag has a huge and disproportionate impact in the wider marine environment.
>
> Zac Goldsmith, British environmentalist, 2019

AN AMAZING NEW PRODUCT

The first plastic was made from cellulose by Alexander Parkes in 1862. In 1869, John Wesley Hyatt took on the challenge from a New York company to find a substitute for elephant ivory. He took cotton cellulose and mixed it with camphor, a type of tree wax. It formed a **flexible** substance that could be made into different shapes. It was thought to be a product that would save the environment because it could replace tortoiseshell, horn, and ivory. People felt it would prevent the hunting and poaching of those wild animals that supplied these materials for household accessories.

For the first time, **manufacturing** was not limited by what was available in nature. There was only so much wood, metal, stone, and bone that could be cut down, mined, quarried, or gathered each year. Now people could create their own material that had many different uses. Plastic was also cheaper than many of those **natural resources**. This meant more people could afford these new products.

NEW SYNTHETICS

In 1907, a new plastic was invented that contained no natural elements in it at all. It was called Bakelite. It could not only be molded into any shape, but was also light, very strong, and, most importantly, heat resistant. This made it useful for the electric lines being wired into homes as electricity became more common. Soon companies were inventing a wide variety of new types of synthetic plastics.

Plastic became even more widespread during World War II. Plastic allowed countries to make more military equipment even when natural resources were scarce. Nylon is a synthetic silk. During the war it was used for parachutes and ropes. Plexiglass is a clear plastic. It was used for windows in military airplanes. During World War II, U.S. plastic production increased 300 percent.

After the war, people were ready to buy new cars, new homes, and new furniture. Plastic seemed to be the perfect solution to cheaper, safer, longer-lasting products. New designs in everything from furniture to appliances were possible because plastic could be any color and any shape.

In the 1960s, people began to notice plastic waste in the oceans. They began to question just

Look around you. How many items can you see that are made of plastic or have plastic parts? Are all of them easy to replace with other materials? Why or why not?

▶ People are concerned with chemicals leaching from plastic packaging into our food. Researchers are developing containers made from **compostable** plant products.

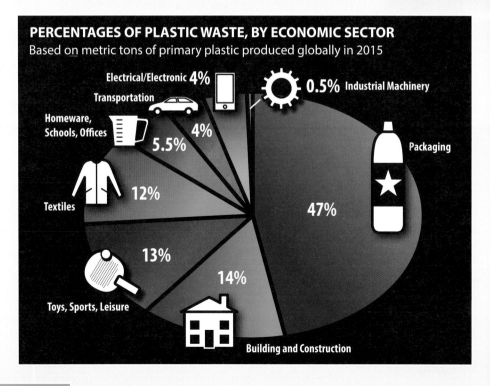

PERCENTAGES OF PLASTIC WASTE, BY ECONOMIC SECTOR
Based on metric tons of primary plastic produced globally in 2015

- Electrical/Electronic **4%**
- Transportation **4%**
- Homeware, Schools, Offices **5.5%**
- Textiles **12%**
- Toys, Sports, Leisure **13%**
- Building and Construction **14%**
- Packaging **47%**
- Industrial Machinery **0.5%**

▲ Today's cars contain 260 pounds (118 kilograms) of plastic. This reduces the weight of the vehicle, saving about 3.8 percent in fuel.

how safe plastics were for their health. Oil spills and chemical **pesticide** problems made the public aware of the need to keep our environment clean. In the 1970s and 1980s, plastic was seen as part of the pollution problem. It was piling up in landfills because it didn't break down. The plastics industry suggested recycling as a solution.

SKYROCKETING PRODUCTION

In 1950, the world produced about 2 million tons (1.8 million metric tons) of plastic per year. By 2015, we were producing 360 million tons (327 million metric tons). High-income countries, such as the United States, France, and Germany, tend to produce the most plastic. A 2015 study by Dr. Jenna Jambeck of the Unversity of Georgia, showed that these countries also tend to have the best recycling and waste-management plans, so do not contribute to plastic pollution as much as countries that produce less plastic, such as Peru and India. China is the exception, as it is not only one of the highest plastic-producing countries, but also the worst when it comes to managing that waste.

Scientists estimate that approximately 8.8 million tons (8 million metric tons) of plastic waste enters our oceans each year. Where does it come from? Countries that have ocean coastlines and poor waste management or recycling programs contribute the most to marine plastic pollution.

About 20 percent of ocean plastic comes from the fishing and shipping industries. This includes fishing nets and sunken boats. The rest comes from land-based sources. Plastics are very lightweight. When we dump plastics into landfills some of it gets blown by the wind to sewers. Litter on city streets is also washed down into the sewer systems. This can then make its way into rivers, which carry it to the ocean. Whether it is blown, thrown, or falls on beaches or into the water, land-based sources make up 80 percent of ocean plastic pollution.

▼ A whale shark accidentally eats a plastic bag in a polluted ocean. Scientists have found microplastics in 114 different species of marine animals.

KEY PLAYERS

The Ocean Cleanup is an organization based in the Netherlands that is interested in removing plastic pollution from the oceans. Boyan Slat, an 18 year old from Delft, is the Dutch inventor who started the group. The Ocean Cleanup uses barriers to surround floating garbage patches so the waste can be scooped up. After some design problems, tests done in October 2019 have shown its system is collecting and holding plastic waste and even microplastics.

TOTAL MISMANAGED PLASTIC WASTE
Quantity that enters oceans

Country	Millions of metric tons (mostly for 2010)
China	8.8
Indonesia	3.2
Phillippines	1.9
Vietnam	1.8
Sri Lanka	1.6
Egypt	1.0
Thailand	1.0
Malaysia	0.9
Nigeria	0.9
Bangladesh	0.8
Brazil	0.5
United States	0.3

0 1 2 3 4 5 6 7 8 9
Millions of metric tons (mostly for 2010)

A *Wall Street Journal* report of 2018 showed that low-income and least developed countries have the poorest record of plastic management.

◀ Microplastic beads and flakes found in facial scrubs—here magnified many thousands of times—block digestive and respiratory tracts in marine animals, causing them to **suffocate** or starve to death.

PLASTICS IN OUR FOOD CHAIN

What happens to the plastic in our oceans? Fish and other creatures eat some of it, thinking it is food. Then we eat the fish. Therefore, we are consuming our own plastic waste. Some plastic, such as fishing nets, plastic rings, and fishing line, ends up wrapped around sea life. This can cause injury or death. Because plastic floats, tides and currents gather it into large garbage patches on the surface of the water. Sunlight and waves are breaking the plastic down into small particles called microplastics. The plastic isn't decomposing into its natural elements; it is just in smaller pieces. These microplastics may end up inside our seafood, or in the **sediment** on the ocean floor.

People concerned with the damage to the environment know that plastics do not just cause pollution as waste and litter. Extracting the natural resources needed to make plastics also creates problems for the environment. Producing plastics uses chemicals and energy. Even recycling contributes to polluting our world.

The polymers used to make today's plastics are created using carbon atoms from fossil fuels found deep in the ground. Large drilling machines make holes down to underground **reservoirs** of petroleum. Pumps push air down into the holes so that the petroleum rises to the surface. Accidents and explosions can lead to spills. Petroleum spills can kill animals by coating their feathers or fur, making it impossible for them to move or breathe. In ocean spills, it kills fish and coral reefs.

▶ A staggering 91 percent of all plastic created is not recycled and ends up in landfills or our oceans. Landfills are filling up three times faster than they would without plastic waste.

LOW RISK   HIGH RISK

WHAT'S AT STAKE?

Microplastics are often too small to be seen easily with the naked eye. Do you think people would be more concerned about microplastics in their food chain if they could see them? Would you eat a fish if you could see the microplastics in its body?

▶ The 2016 movie *A Plastic Ocean* details the plastic waste journalist Craig Leeson found while searching for the blue whale.

"The most important film of our time"
Sir David Attenborough

A PLASTIC OCEAN

Proudly Presented by

Education Partner

A PLASTIC OCEAN FOUNDATION

PLASTIC OCEANS LIMITED
presents in association with
ADESSIUM FOUNDATION HEMERA FOUNDATION
A PLASTIC OCEAN
Executive Producers: Sonja Norman Daniel Auerbach
Produced by: Adam Leipzig Jo Ruxton
Directed by: Craig Leeson

PLASTICS AND CLIMATE CHANGE

Changing fossil fuels into polymers for plastics uses chemicals that are thought to cause cancer. **Exhaust** from manufacturing plants and byproducts from the equipment also **contaminate** our air and water. **Phthalates** are chemicals that are added to some plastics to make them softer and easier to bend. Phthalates can evaporate into the air and cause problems when we breathe them in.

Recycling plants need energy to run. They add carbon **emissions** into the air that are a factor in climate change. Even plastics filling up landfills are not harmless. Over time, chemicals in plastics can leach into the soil and water. Bisphenol A is known to people as BPA. It was used in water bottles and food containers but found to cause health problems. Even though few plastics contain BPA now, older plastics in landfills will contain the harmful chemical, and it can still leach into our water supplies underground.

As plastics don't begin to break down for many hundreds of years, the plastics industry suggested recycling as a way to keep plastics out of landfills and oceans. But recycling is not a perfect solution to plastic pollution.

There are two main methods of recycling plastic. One way is to shred the plastic into tiny pieces, melt it down, and reuse it. The problem with this method is that new material has to be added to the recycled pieces to keep them strong and flexible. The other method is to break plastic down into its original elements using chemicals. Then the original elements can be recombined to make new plastic. This produces byproducts from the chemicals that can be harmful. More than half the recycling companies today use the chemical method. With either method, plastic can be recycled only two or three times before it is too **degraded** to be of any use.

THE CENTRAL ISSUES

Many people believe that recycling is the answer to our plastics problem, but recycling uses energy and creates its own waste products. Should we invest more money in developing better recycling methods for more types of plastics or focus on developing alternatives to plastic?

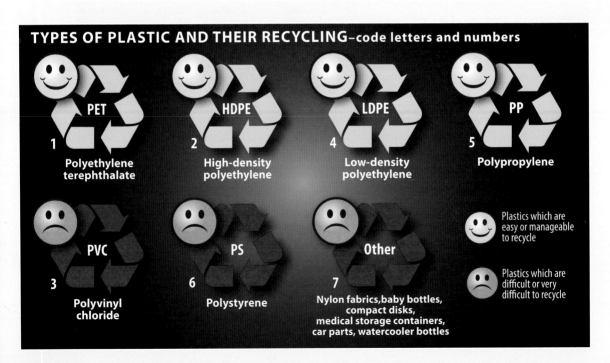

TYPES OF PLASTIC AND THEIR RECYCLING–code letters and numbers

PET 1 — Polyethylene terephthalate

HDPE 2 — High-density polyethylene

LDPE 4 — Low-density polyethylene

PP 5 — Polypropylene

PVC 3 — Polyvinyl chloride

PS 6 — Polystyrene

Other 7 — Nylon fabrics, baby bottles, compact disks, medical storage containers, car parts, watercooler bottles

Plastics which are easy or manageable to recycle

Plastics which are difficult or very difficult to recycle

On the sign:

MACREBUR
The plastic road company

THIS ROAD IS MADE FROM
WASTE PLASTICS
W W W . M A C R E B U R . C O M

▲ Plastic-waste roads sound like a great idea to use discarded plastics but some researchers worry that they will give off harmful gases as the roads are heated by sunlight.

RECYCLING ISSUES

While recycling is the best way to dispose of plastics, how much ends up being recycled? Studies show that in Canada less than 11 percent actually gets recycled into new plastic products. The rest is either **incinerated** or put into landfills. When plastics are incinerated, acids and chemicals in the plastic enter the air. Scrubbers in the chimneys clean the air before releasing it, but the burning and scrubbing produces a very toxic ash. This ash has to be disposed of in **hazardous** waste **facilities** because it is more harmful than the original plastic!

Not all plastics are recyclable. Some plastics are made of extra components that cannot be broken down by either method. Some plastics can be recycled but there might be no market for the recycled product, so instead it is incinerated or sent to a landfill. Sometimes plastics are not recycled because the country or region that produces or uses the plastic product does not have the right facilities or equipment to do it.

4 SUSPENDING JUDGMENT

The issue of our use and misuse of plastics is a complicated one. It affects many parts of society, including businesses, our **economy**, recyclers, and consumers. To really understand what is working, what is not, and what we need to do to keep ourselves and our world safe and healthy means investigating all sides. Seeing different viewpoints gives us a balanced view.

▶ Programs such as The Great Canadian Shoreline Cleanup and the Great Plastic Pick Up in England, U.K., aim to reduce loose plastic garbage and the amount of plastic blowing or flowing into our rivers and oceans.

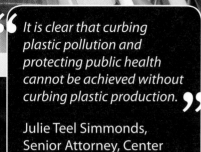

AN ECONOMIC BENEFIT

The production, disposal, and recycling of plastics is a huge industry. It provides about 82,000 jobs in Canada. In the United States, the plastics industry employs almost one million people in manufacturing alone, with more than 1.4 million people working in a job related to plastics.

Some people are concerned that reducing our use of plastics will have a devastating effect on the economy and will lead to the loss of many thousands of jobs. Others argue that the new jobs to replace any lost would be created in new "green" industries. Green industries use environmentally friendly sustainable methods and resources to produce goods and services.

THE PETROLEUM INDUSTRY REACTS

As our use of fossil fuels for energy decreases, oil companies are looking to plastics production to make up the difference. The Plastics Industry Association (PIA) in the United States includes companies such as Exxon Mobil, Shell Polymers, and DowDuPont. The Association has been successful in lobbying governments to strike down laws that would limit the use of plastics.

Studies have shown that charging a small amount for plastic shopping bags has greatly reduced the number of bags ending up in landfills and in our oceans. But this puts the onus on consumers, not on plastic manufacturers. In March 2019, the government of Tennessee, supported by the PIA, signed a bill forbidding local governments in the state from banning plastic bags.

THE CENTRAL ISSUES

Plastics used in health care are vital to keep germs from spreading through contaminated equipment in a cost-effective way. What might we use to replace them, but still keep patients healthy and costs down?

▶ Grocery stores are full of plastic, from soda bottles to plastic wrap around meat and plastic bags to carry produce. New stores are starting to offer plastic-free shopping by using refillable containers and paper bags.

GLOBAL PLASTICS PRODUCTION

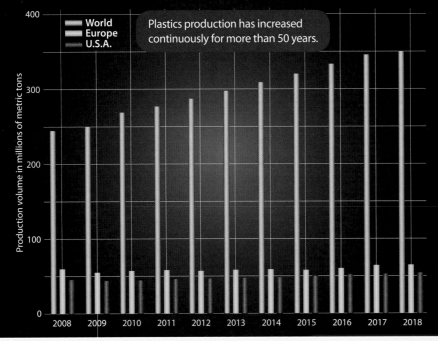

Plastics production has increased continuously for more than 50 years.

Legend:
- World
- Europe
- U.S.A.

Production volume in millions of metric tons

Years: 2008, 2009, 2010, 2011, 2012, 2013, 2014, 2015, 2016, 2017, 2018

Many businesses rely on plastics to keep down their costs. Airplanes made with **reinforced** plastics can lower operating costs and fuel by 15 percent. This means less pollution and **emissions** that contribute to climate change. Lighter-weight cars also use less fuel and are cheaper to transport from the factory to consumers. Many products are packaged using plastic. This means a big savings for companies who ship items over long distances. With heavier packaging, such as wood or metal, increased transportation costs would have to be added to the price, making products too expensive for some consumers to afford.

Some businesses are already looking for alternatives. Restaurants that use plastic straws are trying out paper, rice, or sugarcane straws. Norwegian Cruise Line is also banning plastic straws on all of its 26 ships, eliminating more than 50 million straws every year. Beer brewers, such as Guinness in Ireland, are switching from plastic rings and shrink wrap to 100 percent **biodegradable** or recyclable cardboard.

PLASTICS AND HEALTH CARE

Hospitals and clinics depend on plastics to deliver safe and clean health care. In the past, syringes were made of glass. They broke easily and were difficult to **sterilize** properly. Disposable plastic syringes prevent passing germs between patients. Blood bags, tubing in dialysis machines, and pill casings are all made of plastic. Even prosthetic arms and legs are lighter, more flexible, and more comfortable if made from plastic. Some new inventions, such as heart valves, don't have an older version made of a different material.

Environmentalists argue that a lot of the plastic waste in health care can and should be recycled or replaced with other materials. Before an operation, sterilized tools are covered in a blue plastic wrap to prevent contamination. The wrap is removed before the patient is even in the room, so it does not come in contact with fluids or germs. This could be collected and recycled. Some hospitals are experimenting with a reusable sterilized container instead.

◄ Petroleum refineries are a major source of air pollution because they release toxic chemicals during their processes.

PARALLEL EMERGENCIES

For those who are concerned with the environment, our dependency on plastics is a dire problem. Groups like the Plastic Pollution Coalition say that "plastic pollution and climate change are parallel global emergencies." Because plastic is made from petroleum, we cannot solve one problem without solving the other. As long as we use plastic we are contributing to climate change.

One of the biggest concerns is the growing amount of plastic waste in our landfills and in our water. Scientists cannot determine exactly how long, if ever, plastics take to decompose but estimate it would take hundreds of years. At the rate plastics are piling up, we will eventually bury our planet in plastic. Since plastic was invented, about 9.1 billion tons (8.3 billion metric tons) of it

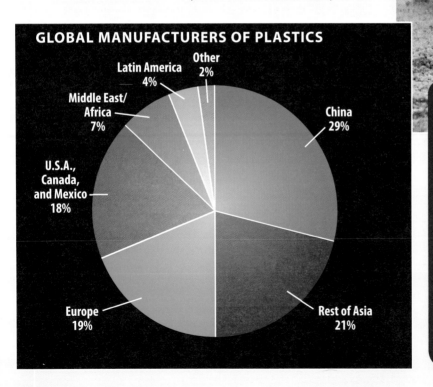

GLOBAL MANUFACTURERS OF PLASTICS

- Other 2%
- Latin America 4%
- Middle East/Africa 7%
- U.S.A., Canada, and Mexico 18%
- Europe 19%
- Rest of Asia 21%
- China 29%

▲ The Food and Agriculture Organization of the United Nations estimates that about 700,000 tons (635,029 metric tons) of fishing equipment is lost at sea each year. When it was made of natural materials, it would decompose. Now most of it is made of plastic, such as plastic nets which entangle and kill ocean life.

have been created. That is as heavy as 1.5 billion elephants! Half of it was made in the last 13 years. Only about 9 percent of this plastic has been recycled in North America. Europe is doing better, recycling 30 percent. Right now there is enough plastic waste in our world to cover a country the size of Argentina. Environmentalists want to improve the amount of plastic that is recycled, and make sure it is recycled more than once. It would be best if plastic was recycled endlessly in a loop.

NO SAFE PLACE

The problem with plastic in landfills isn't just the size. Plastics and the chemicals that may leach from them are polluting our land and water. This is killing animals, who also accidentally eat plastics and microplastics. The pollution is also harming us as we take in the chemicals in our food and water. Chemicals such as polystyrene, which is used in food containers, can lead to immune system problems and even cancer. Environmental groups are working to educate the public on the dangers of plastics and also to get action from industries and governments.

◄ The top ten types of beach litter are all plastic. They include food wrappers, bottles, bottle caps, and grocery bags.

Most communities in North America have a plastic recycling or blue box waste program. Many of them do their part to help the environment. But the recycling industry is struggling. From the industry's point of view, plastics are difficult to collect, sort, and clean. They are costly to recycle and places to sell recycled plastic are disappearing.

For many years, Canada and the United States were selling their plastics to China to be recycled there. North Americans believed it was being turned into new products. We now know that only valuable materials were taken out and the rest was burned. Burning plastics puts toxic chemicals into the atmosphere. Now, China is taking very little plastic waste. Other countries, such as India, Taiwan, and South Korea, are limiting what plastic waste they will take. North America has to take responsibility for the amount of plastic it uses and throws away.

A FAILING IDEA

Recycling programs are costing taxpayers a lot of money with little benefit because much of the plastic collected in the blue boxes is ending up in landfills. We need a new solution. The province of British Columbia has a new law: Anyone who produces or **imports** a product that will be collected in a blue bin has to pay for its recycling. Now, 69 percent of products in the province are recycled, the highest rate in North America. The fees to companies are lower for materials that are easily recycled: Paper egg cartons cost companies less than Styrofoam ones, for example.

The Canadian government hopes this will get companies to make changes to their packaging. Environmental groups want to create a recycling program so more regions find a healthy solution to the growing amount of plastic going to landfills.

▼ The first step in the recycling process after collection is removing nonplastic items, then sorting the plastics by type and color.

HOW IS PLASTIC PACKAGING WASTE DEALT WITH IN DIFFERENT COUNTRIES?

Percentage

| | 100 | 75 | 50 | 25 | 0 | 25 | 50 | 75 | 100 |

Croatia
Romania
Hungary
UK
France
Lithuania
Italy
Netherlands
Norway
Germany
Austria
U.S.A.
Canada

■ Wasted: put in landfill   ■ Recycled   ■ Reused: burned for energy

▲ Plastic bottles in these bales are ready for recycling. They can be turned into carpets, backpacks, and sleeping bags.

LOW RISK   HIGH RISK

WHAT'S AT STAKE?

Should we keep expanding recycling programs and building new plants even if recycling itself uses a lot of energy and produces pollution? Should taxpayers continue to fund these programs if they are not helping the environment?

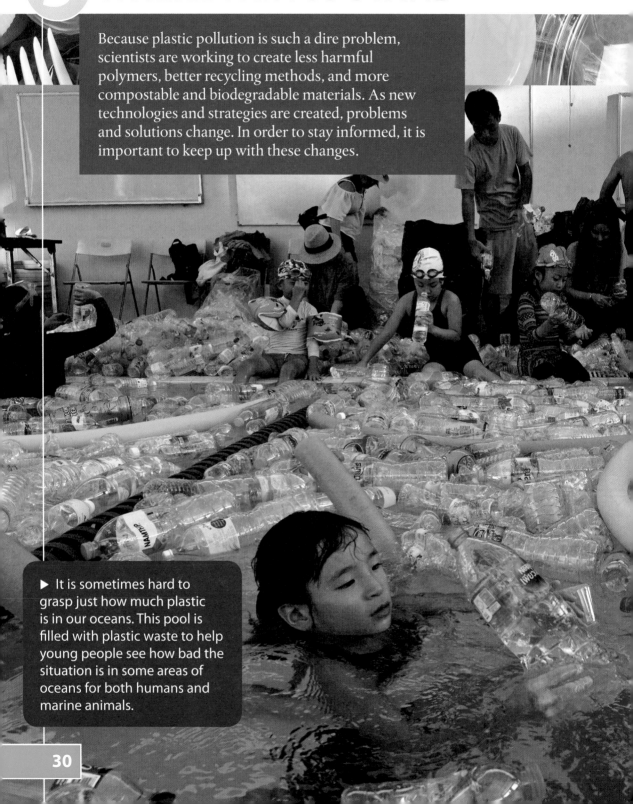

Because plastic pollution is such a dire problem, scientists are working to create less harmful polymers, better recycling methods, and more compostable and biodegradable materials. As new technologies and strategies are created, problems and solutions change. In order to stay informed, it is important to keep up with these changes.

▶ It is sometimes hard to grasp just how much plastic is in our oceans. This pool is filled with plastic waste to help young people see how bad the situation is in some areas of oceans for both humans and marine animals.

THE CENTRAL ISSUES

How important is research into new types of plastic? Should we focus on plastics that can be easily, cheaply, and endlessly recycled? Or plastics that are biodegradable or compostable?

> *Getting rid of plastics entirely is highly unlikely, but also unnecessary. What we need to do is learn to stop using bad plastics and start using good plastics instead.*
>
> Professor Anthony Ryan, University of Sheffield, 2018

SPREADING MISINFORMATION

Newspapers, radio, websites, and social media are all ways that information is shared. But misinformation can be spread as quickly and easily as accurate information. Each time misinformation is heard or read, the more likely someone is to believe it. It is important to always check your sources and dig deeper to find out if something is true. Check out the identity of the person or group giving information and see if claims are backed up with scientific research.

SHOWING TWO FACES

Misinformation is sometimes passed on by accident. Sometimes it is done on purpose to promote one view or focus on an issue. Companies in the plastics industry depend on their business growing to keep profits high. Their goal is to keep producing and selling plastic items. Their bias is **promoting** how safe and useful their products are.

Some organizations are promoting recycling with one hand, while trying to stop **restrictions** on plastic use with the other. The American Progressive Bag Alliance sponsored a contest called A Bag's Life to encourage kids to spread the word about cleaning up plastic waste. At the same time, they supported the Tennessee government's bill to stop local governments from banning plastic bags. Some producers also try to get around laws designed to protect the environment by dumping plastic in countries without good waste and landfill programs, knowing that most of the plastic will end up polluting the environment.

Eliminating all plastics from every part of our lives is not possible at the present time. Some plastics, such as medical supplies, airplane panels, and electronic parts, are not easily replaced with another material. Many people believe though that we can make a start in reducing plastic waste and pollution by targeting single-use plastic.

Single-use plastics are the items we use once and then throw away. The food industry uses many types of single-use plastic, including straws, cutlery, Styrofoam cups and plates, plastic carrying bags, and plastic wrap. This type of plastic is not easily recycled. It is also lightweight, making it easy for the wind to pick it up and deposit it in waterways or the oceans.

Learning to use less single-use plastic and to dispose of it properly is a good thing. But researchers say the answer is not as easy as just switching to cotton or paper bags. They say we have to also look at the impact on the

NEED A STRAW?
AS of July 22nd we have transitioned to **Strawless lids.**
Straws will only be given out with Frappuccinos® and beverages with whipped cream. If you need a straw...
Just ask your barista we are happy to help!

▲ Eliminating plastic straws is a first step for many companies looking to cut down single-use plastic. The new materials developed for compostable straws might lead to compostable utensils, plates, and take-out food containers.

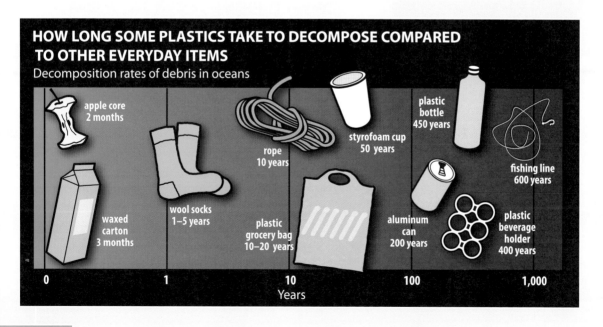

HOW LONG SOME PLASTICS TAKE TO DECOMPOSE COMPARED TO OTHER EVERYDAY ITEMS
Decomposition rates of debris in oceans

apple core
2 months

rope
10 years

styrofoam cup
50 years

plastic
bottle
450 years

fishing line
600 years

waxed
carton
3 months

wool socks
1–5 years

plastic
grocery bag
10–20 years

aluminum
can
200 years

plastic
beverage
holder
400 years

| 0 | 1 | 10 | 100 | 1,000 |

Years

NO TO SINGLE USE PLASTICS

Nusantara   EcoNusa   WALHI   GREENPEACE
                        JAKARTA
              FILM   kalā
              NASIONAL

▲ Protestors in Jakarta, Indonesia, created a 46-foot (14 m) tall plastic fish made entirely from various plastic wastes to bring awareness to the amount of plastic waste threatening ocean life.

environment of producing items made of different materials. A study in Denmark found that the production of paper and fabric bags uses much more land and water, and gives off much more carbon dioxide and other emissions, than the production of plastic bags.

PLASTIC FREE

Many companies are beginning to look at ways to reduce their use of plastic. They are advertising their goals to be "Plastic free by 2025." Nestlé plans to make 100 percent of its packaging recyclable or reusable by this date. Worldwide retail store IKEA plans to ban single-use plastic in all parts

of its stores, including its restaurants and packaging, even earlier, by 2020. Starbucks also plans to ban plastic straws by 2020 and offers a discount to customers who bring their own reusable cups. Many customers are also willing to trade some inconvenience and slightly higher prices to protect the environment. But banning plastics alone will not solve everything. Inventing new materials and adopting new behaviors are also part of the solution.

Myths are pieces of misinformation that have been shared between people so often that they are accepted as true without people feeling the need to double-check the facts. Myths prevent us from questioning the information we receive. Sometimes myths are spread on purpose to support one side or the other of an issue.

Some myths about our dependency on plastic are:

- **Not all plastics are recyclable.**

All types of plastic can be recycled (see graphic on page 20). But there may be big hurdles that make the recycling process too difficult or too expensive.

Some recycling methods may require special equipment or chemicals. Some smaller communities may not have the money or the right facilities to recycle those types of plastic, so they will not accept them in their blue boxes. Sometimes products combine different types of plastics and it is too difficult or too expensive to sort them out and separate them, so they are sent to a landfill.

- **Bioplastics will eliminate waste issues.**

Bioplastics are plastics made with **organic** material. But these are still artificial materials and they will take a really long time to decompose in landfills. Like other types of plastics, bioplastics can be recycled and this is the best solution to dispose of them.

- **There is a floating island of plastic in the Pacific Ocean.**

Many people talk about a large patch of garbage collected together by ocean currents in the Pacific. Some websites say it is as large as the country of France. People picture it as plastic bottles, bags, and packages floating on the surface. It is actually mostly made up of small, fingernail-sized pieces of plastic making the water look soupy. It cannot be seen by satellites overhead. But it is still an environmental disaster. These pieces are not decomposing but are breaking down into small particles. They often sink down to the ocean floor and get eaten by marine life thinking it is food.

▶ These Cambodian students have turned plastic water bottles into their country's flag hanging on the wall behind them. About one million plastic bottles are bought around the world every minute.

▲ Separating out the plastic from the rest of our garbage and recycling it saves twice as much energy as burning it in an incinerator.

WHAT'S AT STAKE?

LOW RISK HIGH RISK

Myths are often believed because they sound like they could be true and are spread by people we trust such as friends and family. What is the danger of accepting myths as truth without checking the facts?

Some researchers believe the solution to plastic waste is to create new **formulas** for environmentally friendly plastic.

Biodegradable material breaks down naturally over a short amount of time into harmless ingredients. Compostable material also breaks down but only with certain conditions such as heat and moisture. Will biodegradable or compostable plastic solve all our problems? Are they even possible?

Researchers from the University of Plymouth in England did a study of these new plastic formulas. They compared a regular plastic bag, a biodegradable bag, and a compostable bag that had been in the ground for three years. They found that the regular plastic and the biodegradable plastic bags were still intact and strong enough to hold groceries. The compostable bag had broken down more than the other two and ripped when researchers tried to put items in it. But it was still there. The scientists concluded that the formulas that have been developed so far cannot be relied on to break down much faster than before. They suggest that advertising for compostable or biodegradable plastic is misleading and these new formulas are not a solution.

FEEDING INSTEAD OF KILLING

Other companies are turning to other materials to replace plastic. A brewery in Delray Beach, Florida, has come up with an alternative to the plastic rings that hold six beer cans. These plastic rings often end up around the necks, wings, or legs of turtles, birds, or fish. The Saltwater Brewery has created new rings from leftover grain mash from brewing that hardens into a firm material. If these rings end up in water, they decompose completely and feed marine life.

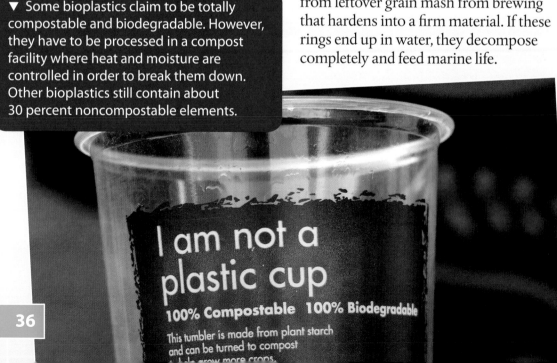

▼ Some bioplastics claim to be totally compostable and biodegradable. However, they have to be processed in a compost facility where heat and moisture are controlled in order to break them down. Other bioplastics still contain about 30 percent noncompostable elements.

I am not a plastic cup

100% Compostable   100% Biodegradable

This tumbler is made from plant starch and can be turned to compost

▶ One ton (0.9 metric tons) of recycled plastic saves the electricity you use in a year, gas for car journeys up to 20,000 miles (32,189 km), gas central heating for two homes for a year, and a roomsize landfill space.

6 KEEPING UP TO DATE

The plastics dependency issue has many sides, each with its own motivation. Even when new technologies are developed, they are not always reported to the public. Petroleum companies may not want people to focus on the growing dangers from plastic in our oceans. Environmentalists may not want news about inventions in biodegradable plastics shared because it might make people think they no longer need to use less plastic.

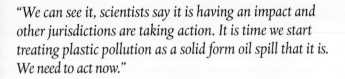

"We can see it, scientists say it is having an impact and other jurisdictions are taking action. It is time we start treating plastic pollution as a solid form oil spill that it is. We need to act now."

This 2019 statement, about discarded plastic along the coast and in the coastal waters of British Columbia by Chloé Dubois, President of the Ocean Legacy Foundation, helps drive forward scientific studies about plastics and their effects on our health and environment. The latest research is investigating how ocean plastic is absorbing toxins that are present in the water before being swallowed by marine life and then entering our own bodies. Other studies are measuring the amount of microplastics building up in beach sand. You should collect and **analyze** such data to make the best decisions about the plastic problem.

YOU NEED TO BE PROACTIVE

Staying informed involves actively looking for source materials and keeping up to date with the latest discoveries and studies. Sometimes you have to ask questions or request material to get to the bottom of an issue. Government agencies, such as Environment and Climate Change Canada or the U.S. Environmental Protection Agency, publish reports that are available to the public. Lists of environmental groups and their contact details can be found online. Plastic industry associations have data on their websites. Newspapers and magazines report on new inventions and pollution disasters online or in print.

When your search and analysis are quick and easy, you are more likely to keep up to date. Ask for help at a library reference desk to find magazines, journals, and books in print form. Librarians can also help you set up a digital borrowing account to read items on your electronic devices at home or on the go.

SEARCHING AND FINDING

News outlets, such as *Global News* or *The New York Times,* can be searched and organized by a single topic. Use the search button and type in your subject such as "Plastic," "Plastic pollution," or something similar. You can also set up an alert for words or phrases on Internet search engines such as Google. This will make sure you don't miss the latest news and developments.

WHAT'S AT STAKE?

What is the benefit of collecting many different sources of information? What is the likely outcome if information is hard to find or difficult to understand?

▼ The Internet is not just a great resource to find information; it is also a great place to connect with people who have the same interests and concerns.

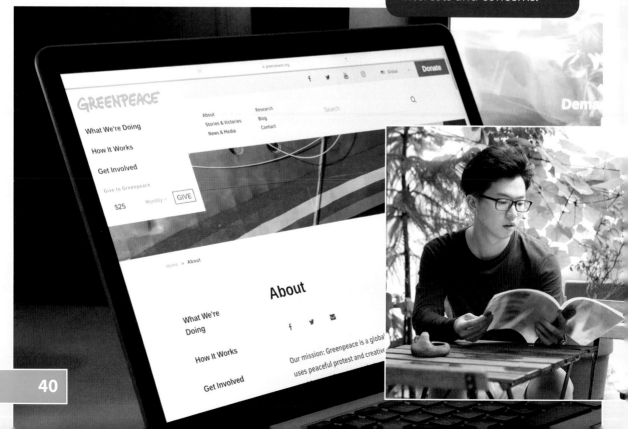

STOP PLASTIC WASTE.
STOP BIG OIL.

▲ New York City mayor Bill de Blasio signed an executive order making city agencies eliminate plastic food containers and cutlery and replace them with compostable or recyclable alternatives.

GET IT TOGETHER

The Internet contains a large variety of websites related to your search on plastics in our world. News aggregator sites collect and organize websites on related issues. Feedly, Google News, SmartNews, and AP News are just some of the aggregators that help you manage your **news diet**. Some aggregators, such as SciURLs, are just for science topics. Others, like TechURLs, let you search the latest headlines on tech subjects. Bioplastics News has the most current news on bioplastics, biodegradable plastics, and bio-based plastics.

On your personal computer, smartphone, or tablet, you can set up a folder of links to your favorite sites and news aggregators. Here you can list all your sources in one spot for easier reference. You can also add links to plastic-free and zero-waste shopping sites that might interest you.

INTERNET SEARCHES

Make your searches quicker and more accurate with these tips:

- Use quotation marks around phrases to search for that exact combination. Searching for plastic pollution will give you sites on all news about plastics and also different types of pollution. Searching for "plastic pollution" will narrow your search to just the pollution that comes from our plastic use.

- Use a colon to limit your search to only specific sites. For government publications on plastics try *Plastics:gov*

The last step in staying informed on your subject is to **audit** your news diet. Auditing means understanding where your news comes from, how accurate it is, and what slant or bias it might contain.

How accurate and reliable your news sources are will depend on who owns, manages, or provides the information. An article on the chemicals found in plastics and how they react to heat, sunlight, or water will give you the most accurate details if it is written by a chemist or other scientist. Information about new laws and regulations facing oil companies are best coming from government websites and reports. The newest programs and clean-up efforts for our oceans will be found in articles and websites of environmental groups. Knowing who has created or reported on the information will allow you to identify the slant or bias, too.

FROM ALL SIDES

Look for different viewpoints on the subject, even if you already have a strong opinion. Seeing a topic from all sides gives you a balanced view. If you want to know more about advances in biodegradable plastics, read articles and reports from **petrochemical** and waste-management companies, recyclers, and environmentalist groups.

Get in the habit of checking facts that you hear. Even the most reliable sources can give false data. Be aware of myths and ask questions if something doesn't sound right to you. If you find facts that seem to contradict one another, dig deeper to see which facts are supported by scientific studies. Read widely to see if there is more support for one view over the others. By verifying sources, you can make the best choices about how and where you use plastics and do your part to keep our world healthy.

INTERNET SEARCH TIPS

When looking at websites, address extensions can help identify the sources of the information.

.gov (government): Official government organizations or departments; you may not be able to access all areas of these websites.

.org: Usually nonprofit organizations and charities; you may have to register to use these.

.com (commercial): Mostly businesses; this is the most widely used web address extension.

Country extensions:
.ca   Canada
.us   United States
.au   Australia
.uk   United Kingdom
.ru   Russia
.de   Germany

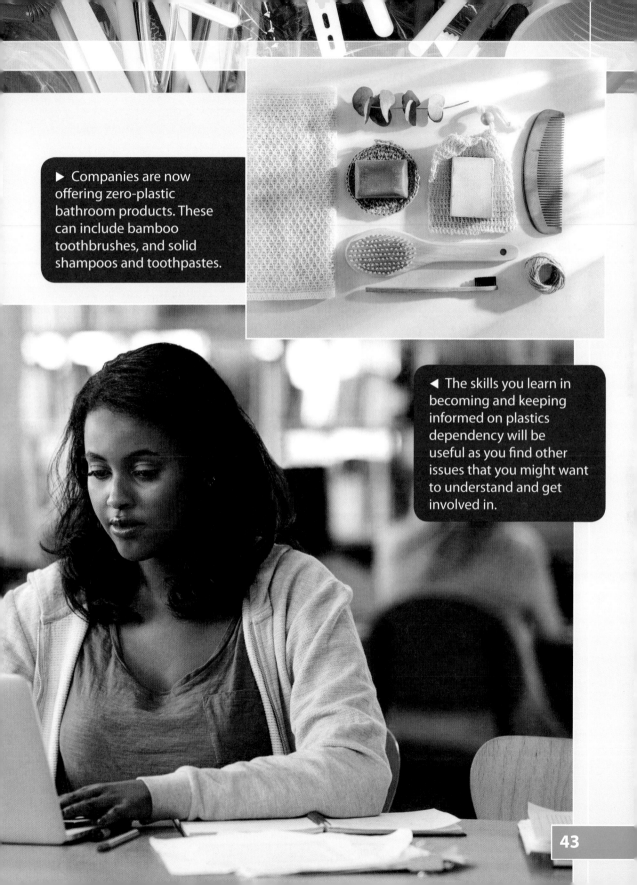

▶ Companies are now offering zero-plastic bathroom products. These can include bamboo toothbrushes, and solid shampoos and toothpastes.

◀ The skills you learn in becoming and keeping informed on plastics dependency will be useful as you find other issues that you might want to understand and get involved in.

GLOSSARY

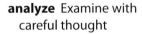

analyze Examine with careful thought

archives Permanently valuable source materials

audit To systematically review or assess something

bias Prejudice in favor of or against one thing, person, or group compared to another

biodegradable Capable of being decomposed by bacteria or other living organisms

bioplastics Biodegradable plastic made from organic substances instead of petroleum

celluloid A transparent plastic

cellulose The fiber that makes up the walls of plant cells

clickbait Content on the Internet that trys to get people to click on a link for a specific web page

compostable Kitchen waste and other garbage that can be broken down and turned into compost

consumers People who buy things

contaminate Make something poisonous or less pure

context The circumstances or setting in which an event happens

convenient Least difficult

daunting Making you feel unsure about achieving something

debate A discussion with different sides

degraded The quality is worse

dependence Relying on something

durable Long-lasting

economy The consumption of goods and services and the supply of money in a country

emissions Substances given off into the air

environment The surroundings in which a person, animal, or plant lives or operates

exhaust Waste gasses expelled from an engine or other machine

facilities Buildings with a specific purpose

fiber A thin thread of natural matter that forms in vegetation

flexible Capable of bending easily without breaking

food chain The series of processes by which food is grown or produced, sold, and eventually consumed

formulas Plans to use ingredients in a certain way

hazardous Risky or dangerous

impacting Having a powerful effect

imports Goods brought in from another country

incinerated Burned completely

interpretations Many different ways to explain something

landfill A hole in the ground where waste material is disposed of by burying it

leaching Dissolving and moving out or away from something

manufacturing Making articles on a large, industrial scale using machinery

microplastics Tiny particles of plastic

motivate Provide someone with a reason for doing something

myth A held but false belief or idea

natural resources Natural materials such as the land, water, minerals, and forests

news diet The sources used to get information

nylon Strong, silky plastic material

organic Belonging to a living thing

pesticide Chemical substance used to destroy organisms harmful to cultivated plants and crops

petrochemical A chemical that comes from petroleum or natural gas

phthalates Chemicals added to some plastics to make them softer and easier to bend

pollution Chemicals that are harmful to the environment

polymer Chemical substance built up mainly or completely from a large number of similar units bonded together

promoting Supporting and encouraging

recycling Converting plastics and other waste materials into reusable form

reinforced Made stronger with another material

reservoir A place where liquid collects

resin A solid or liquid synthetic organic polymer used to make plastics

restrictions Conditions that limit something

sediment Little bits of soil or rock that settle at the bottom of the ocean

source material Original document or other piece of evidence

statistics The collection, analysis, and presentation of numerical data

sterilize Make something free from bacteria

suffocate Die from lack of air or inability to breathe

synthetic Made from artificial substances

toxic Poisonous

vinyl A tough plastic

45

SOURCE NOTES

QUOTATIONS

p. 4: https://bit.ly/324z1Hr
p. 9: https://www.ciel.org/news/
plasticandhealth/
p. 13: https://bit.ly/38fxRfL
p. 23: https://www.ciel.org/news/
plasticandhealth/
p. 31: https://ab.co/2NuRUyl
p. 39: https://bit.ly/2oylCtN

REFERENCES USED FOR THIS BOOK

Chapter 1: The Need to Know, pages 4–7
https://www.theworldcounts.com/stories/
Pollution-from-Plastic
https://www.globalcitizen.org/en/content/
plastic-pollution-facts/
https://bit.ly/326ie6R
https://bit.ly/3338kEl

**Chapter 2: How to Get Informed,
pages 8–11**
https://www.plastics.ca
https://www.psychologytoday.com/us/
basics/bias
https://literarydevices.net/context/
https://bit.ly/2pkrVBE

**Chapter 3: How Did We Get Here?,
pages 12–21**
https://www.sciencehistory.org/the-
history-and-future-of-plastics
https://bit.ly/2NsjwUt
https://njersy.co/2psct6g
https://www.wwf.org.uk/updates/how-
does-plastic-end-ocean
https://theoceancleanup.com
https://mayocl.in/2pu4dTj
https://bit.ly/2N0s9qj

https://bit.ly/2Nv5nGa
https://bit.ly/2r0ZyZB

**Chapter 4: Suspending Judgment,
pages 22–29**
https://bit.ly/34feYaN
https://www.bpf.co.uk/industry/benefits_
of_plastics.aspx
https://bit.ly/2qajSag
https://on.natgeo.com/2WrldFW
https://www.rd.com/culture/companies-
getting-rid-plastic/
https://ecologycenter.org/factsheets/
adverse-health-effects-of-plastics/
https://www.ciel.org/news/
plasticandhealth/
https://on.natgeo.com/2WvFDhd

**Chapter 5: Where Things Stand,
pages 30–37**
https://bit.ly/2JvDA7t
https://bit.ly/34eLwlj
https://www.hefty.com/hefty-energybag/
hefty-energybag-program
https://bbc.in/2NtxSnF
https://bit.ly/2WvUyYC
https://bit.ly/34sgJ4T

**Chapter 6: Keeping Up to Date,
pages 38–43**
https://www.taxpayer.com/media/En4-
366-1-2019-eng.pdf
https://bit.ly/2N2532P
https://bit.ly/335gJY3
https://washedashore.org
https://bioplasticsnews.com
https://bit.ly/2N6gvul

FIND OUT MORE

Finding good source material on the Internet can sometimes be a challenge. When analyzing how reliable the information is, consider these points:

- Who is the author of the page? Is it an expert in the field or a person who experienced the event?

- Is the site well known and up to date? A page that has not been updated for several years probably has out-of-date information.

- Can you verify the facts with another site? Always double-check information.

- Have you checked all possible sites? Don't just look on the first page a search engine provides.

- Remember to try government sites and research papers.

- Have you recorded website addresses and names? Keep this data so you can backtrack later and verify the information you want to use.

WEBSITES

The Maine Department of Environmental Protection lets you see what your recycled material will become:
https://bit.ly/2Nw1il0

Plastic Adrift lets you see where plastic dumped into the ocean will end up in this interactive map:
http://plasticadrift.org

Take the Oceans Plastic Pollution Quiz with Earthday Network:
www.earthday.org/oceans-and-plastic-pollution-quiz/

ABOUT THE AUTHOR
Natalie Hyde has written more than 75 fiction and nonfiction books for kids. She shares her home with a little leopard gecko, and a cat that desperately wants to eat it.

BOOKS

Eriksson, Ann. *Dive in!: Exploring Our Connection with the Ocean*. Orca Book Publishers, 2018.

Fabiny, Sarah, and Dede Putra. *Who Was Rachel Carson?* Grosset & Dunlap, 2014.

French, Jess. *What a Waste: Trash, Recycling, and Protecting Our Planet*. DK Publishing, 2019.

Moser, Elise, and Scot Ritchie. *What Milly Did: The Remarkable Pioneer of Plastics Recycling*. Groundwood Books 2016.

Paul, Miranda, and Elizabeth Zunon. *One Plastic Bag: Isatou Ceesay and the Recycling Women of the Gambia*. Scholastic Inc., 2017.

Salt, Rachel. *The Plastic Problem*. Firefly Books, 2019.

INDEX